# THE PRODUCT DEVELOPMENT PROCESS

## FROM CONCEPT TO COMMERCIALIZATION

A SYSTEMS APPROACH TO BUILDING WHAT MATTERS

IMRAN H. KHAN

# CONTENTS

## Part II

## QUEST

### Translating Intent into Learning

## Part III

## ACCELERATE

From Validated Learning to Disciplined Execution

The Product Development Process: From Concept to Commercialization
A Systems Approach to Building What Matters



Print ISBN: 979-8-9953062-0-7
eBook ISBN: 979-8-9953062-1-4

First Edition, 2026
Printed in the United States of America
Book design by the author/Indie Book Designer
10 9 8 7 6 5 4 3 2 1

# DEDICATION

*To*
*Salahuddin Khan and Micah Drayton, without whom this journey would not have been possible.*

# INTRODUCTION

In 2007, Shai Agassi stood before some of the world's most influential investors and described a vision that seemed inevitable. A former SAP executive named TIME's list of the hundred most influential people, Agassi proposed to free the world from oil dependence by building a global network of battery-swapping stations for electric cars. Drivers would pull in, swap a depleted battery for a charged one in under three minutes, and drive on, unlimited range, zero emissions. It was elegant. It was technically sound. The stations worked exactly as designed.

The world believed him. General Electric, HSBC, Morgan Stanley, and Israeli billionaire Idan Ofer invested. Renault committed to manufacturing a custom electric sedan. Israel and Denmark signed on as launch markets. The company, Better Place, raised nearly eight hundred and fifty million dollars, one of the largest sums ever committed to a clean-technology startup. It became a symbol of Israel's status as the Start-Up Nation.

By May 2013, Better Place filed for bankruptcy. Fewer than fourteen hundred cars had been deployed. The company's assets were liquidated for less than half a million dollars.

Technology had not failed. The battery-swapping stations functioned flawlessly. What failed was everything around technology, the assumptions that had never been examined. Agassi's model required automakers to standardize battery packs across competing brands, but no manufacturer had any incentive to do so. In Israel, where half of all new car sales went to fleet leasing companies, Agassi's model threatened to make those companies obsolete, and they refused to buy his cars. The infrastructure cost half a million dollars per station to build, but the market penetration needed to justify that investment never materialized. Better Place expanded to Denmark, Australia, and China before proving the model worked in a single country.

None of these facts were unknowable. All of them were unexamined. The engineering was brilliant. The system design was fatally incomplete.

Better Place is not an unusual story. It is the default story.

About ninety percent of all startups eventually fail. Seventy-five percent of venture capital-backed startups fail, meaning even significant funding does not dramatically change the odds. Technology startups, the sector that produces the most unicorns, also carry the highest failure rate of any industry, at sixty-three percent within five years. The combined probability of a funded company achieving lasting success is roughly one in two thousand.

These numbers are familiar. What is less familiar is where the fatal decisions are made.

It is almost never at the end. It is almost never in execution. The Standish Group's CHAOS reports, which have tracked project outcomes for decades, consistently find that definitional failure projects that delivered the wrong thing, not the thing wrongly, account for a dominant share of all failures. CB Insights analyzed over one hundred startup post-mortems and found that forty-two percent cited no market need as the primary cause of death. Not poor engineering. Not insufficient capital. The wrong problem, or the right problem misunderstood.

The pattern holds whether the product is software, hardware, a pharmaceutical compound, or a spacecraft. Teams commit to a direction before they understand the terrain. Then they execute with discipline and skill, in the wrong direction. The better they execute, the faster they arrive at the wrong destination. Better Place did not lack talent, vision, or capital. It lacked the upstream discipline to test its most consequential assumptions before eight hundred and fifty million dollars made them irreversible.

## THE INVISIBLE ARCHITECTURE

Every building has two architectures. There is one where you can see, walls, windows, the line of the roof against the sky. And there is the one you cannot, the foundation, the load paths, the decisions about soil and structure that were made before a single visible element existed. If the invisible architecture is wrong, no amount of beauty above ground will save the building.

Product development works the same way.

Before a line of code is written, before a CAD model is opened, before a customer interview is scheduled, a series of choices has already begun shaping the outcome. Which problem is being solved? For whom? Under what constraints? With what assumptions about the market, the technology, the regulatory landscape, and the organization itself? These choices form the invisible architecture of every product effort. They determine what becomes possible later. And they are almost never made deliberately.

Better Place's invisible architecture included an assumption that automakers would standardize battery designs for a startup's benefit, that leasing companies would embrace a model that threatened their existence, and that infrastructure economics would work at low volumes. Each assumption was embedded in the plan from the beginning. Each was treated as a given rather than a hypothesis. By the time reality contradicted them, eight hundred and fifty million dollars had already been committed.

This book is about making those invisible decisions deliberately, before capital and momentum make them irreversible.

## SHIFTING THE ODDS

Market timing, product-market fit, and the quality of the founding team consistently emerge as the strongest predictors of which ventures beat the odds. These are not random variables. They are outcomes of decisions about how problems are framed, how alternatives are evaluated, how systems are modeled, how requirements are captured, and how organizations coordinate under uncertainty.

Each of these decisions can be made well or poorly. Made well, they compound. A clearly framed problem focuses experimentation. Focused experimentation produces validated learning. Validated learning informs architecture. Sound architecture enables reliable execution. Reliable execution builds products that endure. At every stage, the probability of success increases not because the team is luckier, but because the team is making fewer consequential mistakes earlier.

Made poorly, the same decisions compound in the other direction. A vague problem framing leads to scattered experiments. Scattered experiments produce ambiguous signals. Ambiguous signals lead to premature commitments. Premature commitments create architectures that cannot adapt. And the team finds itself executing with precision on a foundation of unexamined assumptions, precisely the trajectory Better Place followed from its first investor presentation to its final bankruptcy filing.

The difference between these two trajectories is not talent, funding, or luck. It is process. Specifically, it is the upstream process, the work that happens before building begins, that most teams either rush through or skip entirely.

This book provides that process.

## THREE PHASES, ONE SYSTEM

Product development is not a single linear process, but a collection of fundamentally different activities that must be sequenced with care. This book organizes those activities into three phases, each designed to maximize learning while minimizing irreversible mistakes.

Spark is about seeing clearly before acting. It is where problems are framed, opportunities and constraints are mapped, the market landscape is understood, and success criteria are established. Spark does not produce a product. It produces something more valuable: the clarity that prevents teams from building the wrong thing.

Quest is about learning deliberately before committing. It is where cross-functional teams form around a shared mission, alternative approaches are evaluated through rigorous trade studies, systems are modeled before they are built, and prototypes serve as thinking tools rather than early products.

Accelerate is about executing with coherence, not just speed. It is where validated learning becomes disciplined commitment, where system models replace intuition, requirements become contracts between teams, and the product system expands to include manufacturing, partnerships, customer support, and lifecycle management.

These three phases are not arbitrary stages. They are a system. Each builds on the one before it. The clarity of Spark informs the experimentation of Quest. The validated learning of Quest grounds the execution of Accelerate. Skip a phase, and the next one inherits unexamined assumptions. Rush a phase, and its output lacks the rigor to support what follows.

## WHAT IS MISSING TODAY

Most of what is taught about product development emphasizes speed, tools, or isolated methods. Lean. Agile. Design Thinking. Rapid prototyping. Each offers real value. But almost none addresses the full complexity of real-world product creation. They focus on execution while skipping the upstream work: framing the problem, understanding context, and sequencing choices.

Education reinforces this gap. Curricula are siloed into disciplines that rarely communicate. Engineers learn to build systems. Designers learn to empathize with users. Business students learn market analysis.

Rarely are these skills integrated in a way that reflects the realities of launching a product.

This book bridges that gap. It does not discard existing methods. It provides the scaffolding that makes them work, the upstream rigor that determines whether Lean is efficient or merely fast, whether Agile is adaptive or merely reactive, whether prototyping is learning or merely building.

## WHO THIS BOOK IS FOR

This book speaks to three audiences, each at a different stage of the journey.

Learners and first-time innovators who need to see the full landscape before committing to a path.

Practitioners moving from intent to execution, who need guidance on translating insights into prototypes, trade studies, system models, and market-ready products.

Leaders and decision-makers responsible for outcomes in startups or established organizations.

Across these stages, the principle is the same: the earlier and more deliberately you frame your work, the more freedom and flexibility you retain later. This is not a book about moving faster. It is a book about moving with intention.

## THE PROMISE

The odds against any new product are severe. This book does not pretend otherwise. What it offers is a systematic way to shift those odds, not through luck or genius, but through disciplined thinking applied at the moments where it matters most.

By the end of this book, you will understand the architecture of product development, visible and invisible. You will know how to frame problems before solving them, how to evaluate alternatives before committing, how to model systems before building them, and how to execute with the coherence that comes from structured understanding rather than improvisation.

You will not eliminate risk. No process can do that. But you will learn to identify it early, manage it deliberately, and prevent it from compounding silently into the kind of failure that eight hundred and fifty million dollars and a team of world-class engineers could not overcome.

---

When the right approach is chosen early, progress compounds. When it is not, even flawless execution struggles to recover lost ground. The work begins here.

---

# PART I
# SPARK

## THE ARCHITECTURE OF UNDERSTANDING

*Framing Problems Before Building Solutions*

# WHAT SPARK IS ABOUT

Spark is about discovering and capturing unmet market needs and pain points. It is the discipline of understanding problems deeply, mapping opportunities systematically, and framing decisions clearly before resources are committed and before assumptions harden into constraints.

Most product failures do not stem from poor execution. They stem from acting on incomplete understanding. Teams build solutions to problems they have not fully examined, for markets they have not carefully studied, with requirements they have not rigorously defined. The result is predictable: wasted effort, missed opportunities, and products that fail to connect with real needs.

## THE SPARK PROCESS

This section of the book is organized around the disciplines that comprise effective early-stage product development. The journey begins with assembling the right group of people, cross-functional teams that bring diverse perspectives to problem framing. It proceeds through opportunity identification, where subject matter experts help identify and validate the problems worth solving.

From there, the process moves to solution ideation, where potential approaches are explored systematically. Model-based systems engineering provides tools for capturing and analyzing complexity. Requirements discipline ensures that insights translate into precise and well understood specifications. Strategic alignment connects all of this work to organizational purpose and market reality.

The tangible output of Spark is a “Proposal” document, a multi-faceted document that captures what has been learned, what is known, what remains uncertain, and what a viable solution must achieve.

But the deeper output is something less visible: a shared understanding among team members, stakeholders, and decision-makers that creates alignment before execution begins.

Spark is not a phase to be rushed through. It is the work that makes all subsequent work more efficient and worthwhile.

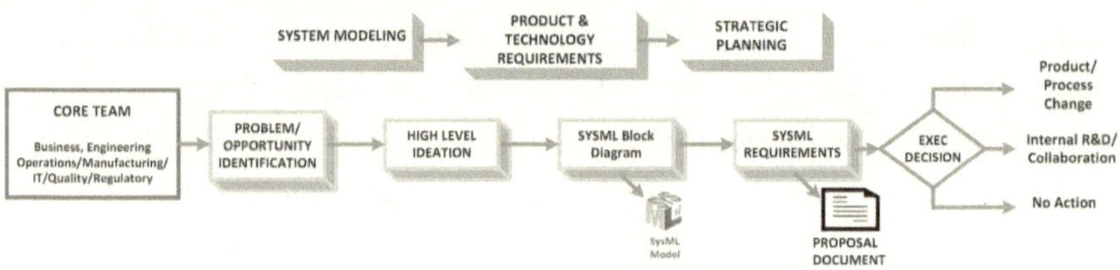


Figure 1: SPARK FLOW

# 1
# ASSEMBLING THE TEAM

The complexity of modern products, spanning technical systems, market dynamics, organizational constraints, and human needs, exceeds what any individual can master alone. Effective problem framing requires diverse perspectives working in concert.

This is why the Spark process begins with team formation. The composition of the team shapes what is seen, what is questioned, and what is ultimately understood.

## THE CASE FOR CROSS-FUNCTIONAL TEAMS

It is important to select people with diverse backgrounds so as to ask questions that are relevant to the problem space being investigated.

Cross-functional composition is about ensuring that the problem is examined through multiple lenses simultaneously: technical feasibility, market viability, operational practicality, and strategic alignment.

## SUBJECT MATTER EXPERTS

Beyond the core team, effective problem framing requires access to subject matter experts, individuals with deep knowledge of specific domains. These might include clinical practitioners for medical products, regulatory specialists for compliance-sensitive industries, or experienced operators who understand how systems actually function in the field.

Subject matter experts serve as a critical function: they ground theoretical discussions in practical reality. They know what has been tried before, what failed and why, and what constraints are truly fixed versus merely habitual. Engaging them early ensures that problem framing reflects how the world actually works, not how the team imagines it might.

The key is selecting experts who bring not just knowledge but intellectual honesty, the willingness to challenge assumptions and surface uncomfortable truths.

## TEAM DYNAMICS

Composition alone does not guarantee effectiveness. Teams must also develop practices that enable productive collaboration across disciplinary boundaries.

This requires psychological safety, the confidence that one can speak candidly without fear of embarrassment or retribution. Amy Edmondson's research at Harvard has shown that psychological safety is the single strongest predictor of team learning and performance. Without it, diverse perspectives remain silent, and the benefits of cross-functional composition never materialize.

Effective teams also establish shared language and shared methods. Disciplinary jargon creates barriers; common frameworks create bridges. The modeling and documentation practices introduced in subsequent chapters serve this function, they provide concrete artifacts around which diverse perspectives can converge.

---

> The team is not just an organizational unit. It is the lens through which the problem will be seen. Choose that lens deliberately.

---

# 2
# SEEING THE PROBLEM CLEARLY

With the right team assembled, the work of understanding begins. Product development starts with the discomfort of engaging the customer in pain. Sometimes the customer needs education about a new way of solving the problem that technology has made available.

The first principle of Spark is simple: clarity precedes action. You cannot design effectively if you do not fully understand the problem, its context, and the constraints that shape what solutions are possible.

## THE COST OF UNCLEAR PROBLEMS

The evidence for rigorous problem framing is overwhelming. The Standish Group's CHAOS reports have consistently shown that definitional failures, building the wrong thing, rather than building the thing wrong account for a substantial proportion of project failures. Research from the Product Development and Management Association suggests that between seventy and ninety-five percent of new products fail to meet their commercial objectives.

Research published by IEEE and other engineering bodies demonstrates that the cost of fixing a defect grows exponentially the later it is discovered. A problem identified during requirements costs roughly one unit to fix. The same problem found during design costs ten to fifteen times more. By the time it reaches testing, the cost has multiplied by fifty. If it survives production, the cost can exceed one hundred times the original.

This pattern holds across industries and project types. The implication is clear: time invested in understanding problems early is among the most leveraged investments a team can make.

## THE QUESTIONS THAT MATTER

Seeing a problem clearly requires asking questions that are often uncomfortable. What is the real need behind the observed behavior? Who experiences this problem, and under what conditions? What assumptions are we making that may not hold?

Consider a company developing a new diagnostic tool. The surface problem might appear straightforward: clinicians need faster test results. But deeper inquiry reveals layers of complexity. Which clinicians? In what settings? What decisions depend on those results? What happens when results are ambiguous? Each question reshapes the problem space and therefore reshapes what solutions are viable.

## ROOT CAUSE ANALYSIS

Effective problem framing often requires distinguishing between symptoms and root causes. What appears to be the problem may actually be a manifestation of something deeper. Solving symptoms provides temporary relief; solving root causes creates lasting change.

Root cause analysis involves tracing observed problems back to their origins. Techniques like the "Five Whys" provide structured approaches: repeatedly asking why a problem occurs until the fundamental cause is revealed.

## PROBLEMS EXIST IN SYSTEMS

Framing a problem also requires contextual thinking. Problems do not exist in isolation; they exist in systems; technical, social, market, and organizational. Each dimension influences what solutions are viable. Ignoring any of them creates blind spots that will emerge later, often at the most expensive moments.

## A STRUCTURED APPROACH

Spark introduces four methods to approach problem framing deliberately:

**Observation**
Step into the environment where the problem exists. Gather data. Record what actually happens, not what is assumed.

**Mapping**
Identify stakeholders, dependencies, and constraints. Visualize relationships and flows.

**Hypothesis**
State what you believe the core problem is, along with the assumptions and uncertainties behind that belief.

**Validation**
Test your framing before committing to a solution. Seek feedback from stakeholders. Look for evidence that contradicts your assumptions.

---

> A well-framed problem is not a guarantee of success. But a poorly framed one is a guarantee of wasted effort.

---

# 3
# MAPPING OPPORTUNITIES AND CONSTRAINTS

Once the problem is seen clearly, the next step is to map the landscape in which it exists. A problem does not occur in isolation; it exists within a web of stakeholders, technical systems, market forces, and organizational realities.

## THE CONSEQUENCES OF UNMAPPED COMPLEXITY

The Boeing 787 Dreamliner program offers a cautionary tale. Boeing's ambitious project aimed to revolutionize commercial aviation with composite materials and unprecedented fuel efficiency. The technical vision was sound. But the program suffered years of delays and billions of dollars in cost overruns.

A significant portion of these problems traced back not to engineering failures, but to incomplete mapping of the supply chain and dependencies. The lesson is universal: complexity that is not mapped does not disappear. It simply emerges later, often at the worst possible moment.

## STAKEHOLDER ANALYSIS

Mapping begins with stakeholders. Who experiences the problem directly? Who is affected indirectly? Who has influence over potential solutions? Stakeholder mapping should identify not just names and roles, but interests and concerns.

## CONSTRAINT DOCUMENTATION

Next comes constraints. These are the boundaries within which solutions must exist. Technical feasibility, regulatory limits, budgetary restrictions, time horizons, and organizational capabilities all form the invisible walls of the problem space. Constraints are not obstacles to be resented; they are tools for focusing effort.

## DEPENDENCY MAPPING

Dependencies are equally important. Understanding what relies on what systems, processes, or people, prevents false assumptions and downstream frustration. Mapping dependencies early reduces the risk of wasted work and unanticipated complexity.

## OPPORTUNITY IDENTIFICATION

Opportunity mapping runs parallel to constraint mapping. Where do small changes create outsized impact? Where does unmet needs intersect with technical or organizational capabilities? Not all problems are equally worth solving. Some offer transformative potential; others offer marginal improvement.

## VISUAL REPRESENTATION

Maps are most useful when they are visual. Block diagrams, relationship charts, and system models make complexity comprehensible. At this stage, precision matters more than speed. The goal is actionable understanding, not execution.

---

> A map is only as good as the thought behind it. The act of mapping is the act of thinking. Clarity in the early stage produces leverage for all that follows.

---

# 4
# UNDERSTANDING THE MARKET LANDSCAPE

A clearly framed problem and a well-mapped landscape are necessary, but they are not sufficient. Problems exist within market dynamic systems of users, competitors, regulation, and trends. To act wisely, you must understand that system.

## MARKET INSIGHT AS COMPETITIVE ADVANTAGE

Consider the Sony Walkman, launched in 1979. The technology was not revolutionary, portable cassette players already existed. What Sony understood was the market landscape: a generation of young people seeking personal, mobile music experiences. The insight was not technical; it was contextual.

## THE REALITY OF MARKET NEED

A 2020 study by CB Insights analyzed over one hundred startup post-mortems. The top reason for failure, cited in forty-two percent of cases, was no market need. These were not poorly executed products; they were products built for markets that did not exist in the form assumed.

## KNOW YOUR CUSTOMER

Market understanding begins with users. Users rarely articulate the problem clearly. They describe symptoms, not causes. They request features, not outcomes. Your role is to translate observation into structured insight.

## WHERE TO PLAY

Strategic market analysis involves understanding where to compete. What ill-served niches exist? What geographies are underserved? What competitors are arriving or departing? What demographic shifts are reshaping demand?

## HOW TO WIN

Teams must understand how to win, what distinctive capabilities or positioning will create competitive advantage. The goal is not to compete in every dimension, but to identify where meaningful differentiation is possible and defensible.

## SYNTHESIS

Understanding the market requires synthesis. Data alone does not produce insight. Without synthesis, information is noise; with it, the system becomes navigable.

---

Clarity of the market landscape turns intuition into informed judgment. Understanding context before designing solutions is the difference between building what is possible and building what matters.

---

# 5
# SOLUTION IDEATION

With a clear understanding of the problem, its context, and its market, the team is ready to explore potential solutions. Solution ideation is the bridge between understanding and action, the process of generating, evaluating, and refining approaches before committing resources.

## FROM PROBLEM TO POSSIBILITY

Ideation builds on everything that came before. Effective ideation is neither purely creative nor purely analytical. It requires the freedom to imagine unconventional approaches and the discipline to evaluate them rigorously.

## STRUCTURED CREATIVITY

Contrary to popular belief, creativity benefits from structure. Studies of innovation consistently show that constraints often enhance rather than limit creative output. The key is the right kind of constraints, those that focus attention on the essential challenge while leaving room for novel approaches.

## INVESTIGATING OPTIONS

Ideation should explore multiple paths, not converge prematurely on a single solution. The investigation should include practical considerations: cost-benefit analysis, build-versus-buy options, evaluation of existing solutions, and exploration of what prototypes might reveal.

## EVALUATION CRITERIA

Ideas must be evaluated against criteria derived from the problem framing and market analysis. Does this approach address the root cause or merely the symptoms? Does it fit within identified constraints? Does it leverage identified opportunities?

## FROM IDEAS TO CONCEPTS

The output of ideation is not a single chosen solution, but a set of viable concepts worthy of further development. These concepts should be documented clearly: what problems they solve, how they work, what assumptions they depend on, and what uncertainties remain.

---

> Ideation is not about generating most ideas. It is about generating the right ideas, those that emerge from deep understanding and point toward genuine solutions.

---

# 6
# MODEL-BASED SYSTEMS ENGINEERING

As concepts emerge from ideation, they must be captured in forms that enable rigorous analysis and clear communication. Traditional prose documents have limitations: they can be ambiguous, they struggle to show relationships clearly, and they are difficult to keep consistent. Model-based approaches address these limitations.

## WHY MODELS MATTER

A model is a structured representation of a system; its components, their relationships, their behaviors, and their constraints. Unlike prose descriptions, models make structure explicit. The Systems Modeling Language, SysML, provides a standardized notation for capturing system architecture, requirements, behaviors, and constraints.

## BLOCK DIAGRAMS

Block diagrams are a fundamental modeling tool. They show the components of a system and how they relate to each other. A well-constructed block diagram makes the system structure visible immediately. During the Spark phase, block diagrams capture the emerging understanding of how a potential solution might be structured.

## REQUIREMENTS DIAGRAMS

SysML also provides notation for capturing and relating requirements. This relational view of requirements is more powerful than traditional lists. It ensures traceability, reveals completeness gaps, and enables impact analysis.

## THE MODEL AS COMMUNICATION TOOL

Perhaps the greatest value of models is their role in communication. Different stakeholders see the same model through different lenses, but they see the same structure. This creates common ground that prose cannot achieve.

## MODELING AS THINKING

The act of building a model is itself a form of thinking. It forces precision. It exposes vague assumptions. It reveals inconsistencies that seemed invisible when the ideas existed only in discussion or document form.

---

> Models are not documentation. They are thinking made visible. They enable rigor without sacrificing creativity.

---

# 7
# CAPTURING REQUIREMENTS WITH PRECISION

The work of previous chapters converges in the discipline of requirements capture. Requirements translate understanding into specification. They bridge insight and implementation. Done well, they enable execution. Done poorly, they guarantee confusion.

## THE PURPOSE OF REQUIREMENTS

Requirements serve multiple purposes. They communicate what must be achieved to those who will build, test, and deliver. They create a reference point for evaluating whether the work has succeeded. They preserve decisions so that the rationale is not lost.

## THE REQUIREMENTS HIERARCHY

Requirements flow from general to specific, from need to implementation.

Market Requirements capture what is needed: customer needs, business needs, and the outcomes that matter.

Product Requirements specify how the product will functionally fulfill those needs.

System Requirements break down product requirements into technical specifications.

Test and Validation Requirements define how compliance with other requirements will be verified.

Each level is traced to the levels above and below it. This traceability ensures that implementation decisions connect to user needs.

## THE LANGUAGE OF REQUIREMENTS

Precision in language matters. In specifications, using the word "shall" indicates a binding provision. To state non-binding provisions, use "should" or "may."

---

> Requirements are the contract between understanding and execution. Precision here creates clarity downstream. Vagueness here creates confusion that compounds.

---

# 8
# STRATEGIC ALIGNMENT

The disciplines of previous chapters create understanding. But understanding must connect to action, and action must connect to purpose. Strategic alignment ensures that effort serves intent.

## THE STRATEGIC HIERARCHY

Vision describes how you see the world at some future time in a manner relevant to your work.

Mission describes what you do that will contribute to realizing that vision.

Goals translate vision and mission into concrete targets with defined endpoints and success criteria.

Projects are organized efforts to achieve goals.

Operating plans organize objectives across time periods, monthly, quarterly, and annually.

## THE POWER OF ALIGNMENT

Consider the Apollo program. President Kennedy's goal was specific and time-bound: landing a man on the moon and returning him safely to earth before the decade was out. This goal aligned with a broader vision of American leadership in space and technology. Without such alignment, effort fragments.

---

> Strategic alignment ensures that there is clarity of understanding of how the product aligns with the overarching strategic plan of the company.

---

# 9
# THE PROPOSAL

The tangible output of Spark is a Proposal, a structured document that captures what has been learned and defines what must be achieved. It converts comprehension into practical, detailed instructions.

## PURPOSE OF THE PROPOSAL DOCUMENT

The Proposal document serves as a bridge between understanding and execution. It communicates the problem to be solved, the constraints that apply, the criteria for success, and the requirements that solutions must meet.

## COMPONENTS OF THE PROPOSAL

A complete Proposal includes several essential components:
**Problem Statement and Context:** A clear articulation of the problem being addressed.
**Opportunity Analysis:** A summary of the opportunity landscape, including market dynamics and competitive alternatives.

**Success Criteria:** Specific, measurable indicators of what success looks like.
**Requirements Specification:** The detailed requirements that solutions must satisfy.
**Constraints and Dependencies:** The boundaries within which solutions must operate.
**Assumptions and Uncertainties:** Explicit acknowledgment of what is assumed and what remains unknown.
**Strategic Alignment:** An explanation of how this work connects to broader organizational goals.

## THE SYSTEMS MODEL

Alongside the Proposal, the Spark process produces a systems model, a SysML representation of the problem space, the proposed architecture, and the relationships among requirements. .

## THE PROPOSAL AS LIVING DOCUMENT

The Proposal at the end of Spark is a living document that evolves as understanding deepens. This document provides decision-makers with the information to take the next step with confidence.

---

> The Proposal captures understanding; the process creates capability. Both are essential outputs of Spark.

---

# PART II
# QUEST

## TRANSLATING INTENT INTO LEARNING

# WHAT QUEST IS ABOUT

Spark produced clarity: a well-framed problem, a mapped landscape of opportunities and constraints, an understanding of the market, documented success criteria, and a Proposal that captures intent. The distance between a clearly stated need and a validated solution is vast, and it is filled with choices that determine whether the effort succeeds or stalls.

Quest is the exploratory journey across that distance. It is where cross-functional teams form around a shared mission. Where alternative approaches are evaluated with rigor, not intuition. Where systems are modeled and prototyped before they are built.

If Spark was about seeing the problem clearly, Quest is about building the solution deliberately.

# 1
# FROM INSIGHT TO EXPERIMENTATION

Once a problem is framed, opportunities are mapped, and success criteria established, the work moves from understanding to action. Insight alone does not create value; it must be tested, explored, and refined. This is the purpose of structured experimentation.

The Lean Startup movement, popularized by Steve Blank and Eric Ries, introduced a generation of entrepreneurs to the concept of validated learning. The core insight was simple but powerful: startups are not small versions of large companies. They are organizations designed to search for a repeatable, scalable business model.

The principle of Quest is straightforward: learn deliberately, not haphazardly. Experiments are not random trials or aesthetic exercises. They are systematic probes into reality.

## LEARNING VELOCITY

Experiments are not only about validating ideas; they are about learning faster. In the modern innovation landscape, learning velocity often matters more than the brilliance of any single idea.

The concept of the minimum viable product, or MVP, emerged from this tradition. But the term is often misunderstood. An MVP is not only a minimal product; it is a minimal test. Its purpose is not only to deliver value to customers, but to learn whether a hypothesis is valid.

The idea of the pivot also belongs here. A pivot is not failure; it is an informed response to learning. Pivots are only possible when the original hypotheses were explicit, and the learning was captured systematically.

## DOCUMENTATION AS COMPOUND INTEREST

Methodical experimentation requires careful documentation. Every step, from hypothesis to observation, must be recorded. This ensures that learning compounds over time.

---

> The work of Quest begins here: with the recognition that building is not the goal. Learning is the goal. Building is the instrument.

---

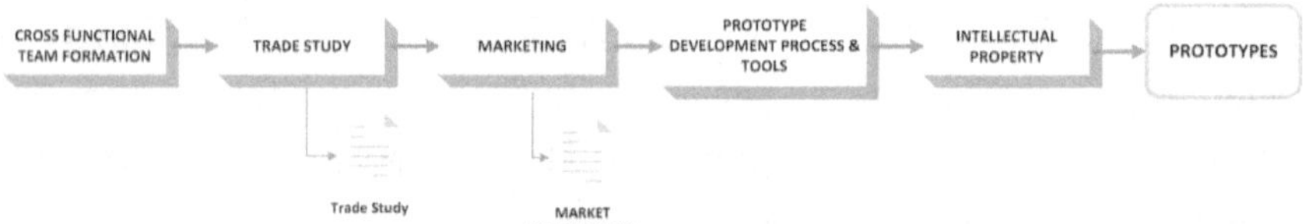


Figure 2: QUEST FLOW

## 2
# THE TRADE STUDY

In the early 2000s, engineers at SpaceX confronted a decision that would define the company's trajectory for the next two decades. The aerospace industry operated on a settled assumption: rockets were assembled from components sourced across a global network of specialized suppliers. Rocketdyne built engines. Honeywell built avionics. Moog built valves. No single company attempted to manufacture the full stack in-house. The supply chain was the architecture.

Elon Musk's team ran the numbers differently. The raw materials in a rocket, aluminum, carbon fiber, copper, Inconel, constituted roughly three percent of the vehicle's market price. The remaining ninety-seven percent was markup, coordination overhead, and inefficiency distributed across a fragmented supply chain. When a supplier quoted a quarter of a million dollars for a single valve, SpaceX engineers estimated they could build it in-house for a fraction of the cost. The question was not whether individual components could be sourced externally. They could. The question was whether the system of interfaces between dozens of suppliers could be managed at the speed and cost the company's mission demanded.

This was a trade study. Not the kind that lives in a spreadsheet, though spreadsheets were certainly involved, but the kind that determines whether an organization will spend the next decade coordinating contractors or controlling its own destiny. SpaceX chose vertical integration, manufacturing roughly eighty-five percent of its launch hardware in-house. The decision collapsed external interfaces into internal engineering conversations, shortened feedback loops from weeks to hours, and eventually reduced launch costs by a factor of ten. Ninety-six Falcon 9 missions flew in a single year.

The SpaceX decision was not obvious at the time. The entire industry's experience argued against it. What made the decision defensible was not conviction alone but the structured analysis that preceded it: the alternatives enumerated, the criteria defined, the trade-offs made visible, and the assumptions documented so they could be revisited as evidence accumulated. That is what a trade study does. It does not eliminate judgment. It elevates it.

## WHAT A TRADE STUDY IS

A trade study is a formal, structured comparison of realistic alternatives against defined criteria. The Federal Aviation Administration's Systems Engineering Manual defines it as the activity of a multidisciplinary team to identify the most balanced technical solution among a set of proposed viable solutions. NASA's systems engineering practice, refined across decades of complex mission development, treats the trade study as a core process, not an optional exercise. For every successful mission, trade studies evaluated propulsion alternatives, communication architectures, material choices, and operational concepts before commitments were made.

What distinguishes a trade study from informal decision-making is documentation and structure. A trade study records the requirements that motivated the decision, the assumptions that shaped it, the criteria by which alternatives were judged, and the priorities assigned to those criteria. This is not bureaucratic overhead. It is insurance. New information arises constantly, a supplier changes terms, a regulation shifts, a technical risk materializes, and decisions must be re-evaluated. A trade study that exists only in someone's memory cannot be revisited. A trade study that is documented can be challenged, refined, and improved.

## THE STRUCTURE OF A TRADE STUDY

At its core, a trade study requires four things: a clearly defined decision to be made, a set of qualified alternatives to evaluate, criteria by which those alternatives will be judged, and a method for weighting and scoring that makes trade-offs explicit.

The criteria emerge from two complementary perspectives. Measures of effectiveness capture the voice of the customer, how well mission objectives are achieved, independent of implementation. Life cycle cost, development schedule, technology readiness, and payload capacity are measures of effectiveness. Measures of performance capture the voice of the engineer, quantitative characteristics of a specific design solution. Mass, power consumption, specific impulse, and throughput are measures of performance. Both matter. Both must be present. A trade study that considers only engineering performance will produce a technically elegant solution that no one can afford. A trade study that considers only cost will produce a cheap solution that does not work.

Weighting is where judgment enters. Not all criteria matter equally. A medical device may weight safety and regulatory compliance far above cost. A consumer electronics product may prioritize time-to-market and user experience. The act of assigning weights forces the team to articulate what matters most and to confront disagreements that might otherwise remain hidden until they surface as conflict during execution.

## SENSITIVITY AND HONESTY

The most disciplined trade studies include a sensitivity analysis: what happens to the ranking if the weights change? If the highest-ranked alternative is sensitive to small changes in weighting, the decision is fragile, it depends on assumptions the team may not be confident about. If the ranking holds across a range of reasonable assumptions, the decision is robust. Sensitivity analysis does not produce certainty. It produces transparency about where certainty is absent.

Trade studies also require intellectual honesty. The temptation to reverse-engineer a preferred outcome, to set weights and criteria that confirm what the team already wants to do, is real and common. Cognitive research has documented the pattern: anchoring bias leads teams to over-weight the first alternative evaluated; confirmation bias leads them to select criteria that favor a predetermined conclusion; group think produces low scoring variance because no one wants to dissent. A trade study that merely validates a predetermined choice is not a trade study. It is a justification dressed in structure.

Guarding against this requires peer review, independent scoring, and a willingness to let the evidence lead. The best teams assign a "red team" role to someone whose job is to challenge the emerging consensus and surface the assumptions everyone else has stopped questioning.

## THE COST OF AN INCOMPLETE TRADE STUDY

The Boeing 787 Dreamliner program again offers what may be the most expensive illustration of this principle in modern industrial history. To develop a revolutionary composite aircraft faster and at lower cost, Boeing outsourced roughly seventy percent of design, engineering, and manufacturing to over fifty partners worldwide. On paper, the trade study favored outsourcing, shared financial risk, access to global expertise, offset deals. But the study modeled the benefits of distributed production without adequately modeling the coordination complexity that distribution would create. Boeing left sub-tier supplier management to its risk-sharing partners, creating a supply chain spanning more than five hundred suppliers in over ten countries. Fuselage sections arrived without wiring or hydraulics installed. A fastener shortage left early airframes held together with temporary bolts. The program, budgeted at five and a half billion dollars and scheduled for five years, ultimately exceeded thirty-two billion dollars and arrived more than three years late.

The lesson is not that outsourcing is wrong. It is that the trade study failed to account for the full system. The variables that were not modeled, organizational interfaces, communication overhead, quality control across cultural and linguistic boundaries, turned out to be the variables that mattered most. Boeing attempted to replicate the benefits of Toyota's supplier relationships without the decades of investment that made Toyota's model work. The trade study evaluated the destination without mapping the terrain.

## SEQUENCING DECISIONS UNDER UNCERTAINTY

Not all decisions carry equal weight, and not all must be made at the same time. A critical discipline in product development is learning to sequence decisions so that the most consequential and least reversible ones are made with the most information, while deferrable ones are held open until context clarifies.

The development of COVID-19 vaccines in 2020 provides a striking counter-example. The traditional vaccine timeline spans a decade. Operation Warp Speed compressed this to under a year, not by eliminating steps, but by resequencing them. Three fundamentally different technology approaches, mRNA, viral vector, and protein subunit, were evaluated simultaneously rather than sequentially. Manufacturing was scaled up during Phase II trials, before efficacy was confirmed, accepting financial risk to compress the timeline. The highest-risk unknowns, biological efficacy and human safety were resolved first, while manufacturing optimization was deferred. This was, in essence, a set of parallel trade studies conducted under extreme time pressure. The structure of evaluation was preserved, every safety threshold was maintained, but the sequencing was redesigned based on a clear understanding of which risks were acceptable and which were not.

## FROM SPREADSHEETS TO SYSTEM MODELS

Traditionally, trade studies live in spreadsheets, static artifacts that capture a decision at a moment in time but lose connection to the system as it evolves. Model-based systems engineering offers an alternative. When trade study criteria, weights, and evaluations are embedded within the system model itself, the trade study becomes a living element of the design. SysML parametric diagrams can encode the relationships between design variables and evaluation criteria, allowing the model to compute scores across alternatives as input parameters change. A spreadsheet trade study is a snapshot. A model-based trade study is a lens that the team can look through repeatedly as the design space evolves.

For early-stage teams, the sophistication of the tooling matters less than the discipline. A structured comparison of three alternatives against five weighted criteria, documented on a single page, can be more valuable than weeks of debate driven by personal preference. The point is not complexity. It is rigor. A clear, documented trade study preserves the reasoning behind a decision, making it available for review, revision, and learning, whether the team uses a spreadsheet, a SysML model, or a whiteboard.

---

> The trade study is where intuition meets evidence. When done well, it does not eliminate judgment, it elevates it. Teams that master this discipline make better choices, faster, and with the confidence that comes not from certainty, but from knowing that the reasoning is sound, the alternatives were considered, and the assumptions are documented for the day when reality demands they be revisited.

---

# 3
# MARKET REQUIREMENTS AND POSITIONING

A product that works is not the same as a product that sells. Technical excellence is necessary but never sufficient. Market requirements capture what the product must achieve within the marketplace to create value and sustain itself.

## THE MARKET REQUIREMENTS DOCUMENT

The market requirements document, or MRD, is a strategic document that captures the relationship between the product and its intended market. A well-crafted MRD includes a clear definition of the target customer, articulates the core value proposition, and establishes measurable success criteria.

Consider the Sony Walkman again. The market requirements, portability, personal use, affordable pricing, and cultural relevance drove product decisions as much as any engineering specification.

## BRIDGING ENGINEERING AND BUSINESS

One of the most common failures in product development is the disconnect between engineering and market perspectives. The discipline of Quest requires that market requirements and product requirements be developed in dialogue, not in sequence.

---

> Market requirements are not an afterthought to engineering. They are the other half of the equation, the half that connects what you build to why it matters.

---

# 4
# PROTOTYPE DEVELOPMENT PROCESS AND TOOLS

In the spring of 2006, a small team at Twitter faced a decision that would shape the trajectory of the product and, arguably, the trajectory of social media itself. The prototype worked. Users could post short status updates. The core interaction was validated. What the team needed next was not more features but a development process that could carry a fragile prototype through the strains of rapid growth without collapsing under its own weight.

They chose Ruby on Rails, a framework optimized for speed of development rather than speed of execution. For a small team iterating on an unproven concept, this was the right choice. Rails let them ship changes in hours instead of days, test ideas against real user behavior, and pivot when the data contradicted their assumptions. The tool matched the moment.

But by 2008, Twitter was experiencing catastrophic failures under load. The infamous "Fail Whale" became a cultural symbol of a service that could not keep up with its own success. The same framework that had enabled rapid prototyping was now the bottleneck preventing reliable operation. Twitter spent the next several years rewriting its infrastructure in Scala and Java, replacing the tool that had been perfect for exploration with tools suited for execution at scale.

The lesson is not that Twitter chose badly. The lesson is that it chose well twice. First, a tool optimized for learning. Then, a tool optimized for performance. The mistake many teams make is choosing once, either selecting tools that are too heavy for exploration or too fragile for scale, and then living with the consequences in both directions.

At some point in Quest, the work shifts from thinking to making. The problem has been framed. Alternatives have been evaluated through trade studies. The market has been understood. Requirements have been captured. Now the team must select and organize the processes, methods, and tools that will carry the product from validated concept to engineered reality. This chapter addresses the infrastructure of that transition: the development process that sequences work, the tools that enable it, and the discipline required to choose both deliberately.

## CHOOSING A DEVELOPMENT PROCESS

There is no single correct development process. The choice depends on the nature of the product, the maturity of the technology, the regulatory environment, and the capabilities of the team. What matters is that the process is chosen deliberately, not inherited by default. A team that has not consciously selected its development process is not following a process. It is improvising.

Waterfall processes, in which work proceeds through sequential phases: requirements, design, implementation, testing, deployment are well-suited to products where requirements are stable, regulatory compliance is critical, and the consequences of late-stage changes are severe. Aerospace, medical devices, and safety-critical systems often follow some form of sequential development, because the cost of iteration at the wrong moment is not merely expensive but potentially catastrophic. When Airbus certifies a flight control system, the regulatory framework demands that requirements be locked before design begins, that design be verified before implementation proceeds, and that each phase produce auditable evidence of compliance. This is not rigidity for its own sake. It is the structure that allows an aircraft carrying three hundred passengers to be trusted with their lives.

Iterative and agile processes, in which work proceeds in short cycles of design, build, test, and learn, are well-suited to products where requirements are uncertain, user feedback is essential, and rapid adaptation creates competitive advantage. Software products, digital services, and consumer applications often benefit from this approach. Spotify's "squad" model, in which small autonomous teams own features end-to-end and release updates every two weeks, allowed the company to iterate its way from a simple music player to a platform that serves over six hundred million users. The process works because the cost of a bad release is low, a software update can be rolled back in minutes and the value of rapid learning is high.

Many products require a hybrid. A medical device may use a structured V-model for verification and validation of its safety-critical firmware while employing agile methods for its companion mobile application. An autonomous vehicle integrates hardware developed on longer cycles with software that evolves weekly. The discipline is not in adhering to a single framework but in understanding which approach applies to which part of the system and ensuring they integrate at defined interfaces.

## THE V-MODEL AND VERIFICATION DISCIPLINE

For products of any significant complexity, the V-model deserves particular attention during Quest. The left side of the V traces the decomposition of requirements: from system-level requirements through subsystem design down to component specifications. The right side traces the corresponding verification: unit tests validate components, integration tests validate subsystems, and system-level tests validate the whole against the original requirements.

The V-model's power is not in its shape but in its insistence on traceability. Every test on the right side is linked to a requirement on the left. Every design decision traces to an intent. This discipline is what distinguishes engineering from tinkering. During Quest, the V-model need not be followed with the rigor it will demand in Accelerate, but the habit of thinking in traced pairs "for every requirement, a verification; for every design choice, a rationale" prevents the accumulation of untested assumptions that will surface later at far greater cost.

The development of COVID-19 vaccines in 2020 offers a striking illustration of how process structure and creative sequencing can coexist. The traditional vaccine development timeline from initial research through Phase I, II, and III clinical trials to regulatory approval typically spans a decade or more. Operation Warp Speed compressed this to under a year, not by eliminating steps, but by resequencing them. Manufacturing was scaled up during Phase II trials, before efficacy was confirmed, accepting financial risk to compress the timeline. Safety and efficacy phases overlapped. The V-model's logical structure was preserved; every verification was still performed but the sequencing was redesigned based on a clear understanding of which risks were acceptable and which were not.

## SIMULATION AS THE NEW PROTOTYPE SHOP

The traditional prototype shop, a physical space where ideas become tangible objects remains essential. But alongside it, a parallel revolution has unfolded. Simulation has become the first prototype for an increasing number of products, and understanding when and how to use it is now a core competency.

Software-in-the-loop testing places software under development into a simulated environment that mimics the physical system it will eventually control. Flight simulators test avionics software against thousands of scenarios that would be dangerous or impossible to replicate in the air. Autonomous vehicle developers simulate millions of miles of driving, generating edge cases, a child chasing a ball into a street, a sudden whiteout on a highway, that no amount of physical road testing could cover economically. Medical device teams simulate patient variability that clinical trials alone cannot capture.

Hardware-in-the-loop testing goes a step further, connecting physical components to simulated environments. A real electronic control unit is wired to a simulator that generates the electrical signals of an engine, a braking system, or an entire vehicle. The hardware behaves as if it is operating in the real world, while the test environment can be controlled, repeated, and stressed in ways the real-world resists. Fault conditions can be injected safely. Stress scenarios can be run overnight. Coverage can be measured systematically.

The value of these approaches is not merely efficiency. It is epistemic. A team that can simulate a thousand scenarios before building a single physical prototype enters the workshop with a fundamentally different understanding of its design space than a team that builds first and discovers second. The simulation does not replace the physical prototype. It makes the physical prototype smarter.

## THE DIGITAL TWIN AS DEVELOPMENT COMPANION

The concept of the digital twin, a dynamic, continuously synchronized digital replica of a physical system, has evolved from an aerospace curiosity into a foundational development tool. During Quest, the digital twin serves a specific purpose: it allows the team to reason about a system that does not yet fully exist in physical form.

A design twin captures the product's architecture, geometry, material properties, and behavioral models in a unified digital environment. Engineers can ask questions before building: what happens when this material is subjected to thermal cycling? How does this assembly respond to vibration at resonance? Where are the stress concentrations in this geometry under worst-case loading? Each question answered digitally is a question that does not need to be answered by a failed physical prototype.

The partnership announced in early 2026 between Dassault Systèmes and NVIDIA illustrates where this trajectory is heading. Dassault's Industry World Models combine four decades of industrial knowledge, physics, engineering laws, system constraints with NVIDIA's accelerated computing platform. The result is what both companies describe as "physical AI": artificial intelligence grounded in the laws of physics, capable of generating plausible future states, evaluating alternative configurations, and enabling engineers to explore design spaces that would take months to traverse through physical experimentation alone. Even NVIDIA itself is now using Dassault's model-based systems engineering tools to design its own AI data center infrastructure, a signal that the complexity of next-generation systems has outstripped what traditional methods can manage.

For early-stage teams, the full digital twin stack may be beyond reach. But the principal scales down. A Simulink model that captures the control dynamics of an embedded system. A thermal simulation in ANSYS that validates a heat sink design before machining. A finite element model that predicts where a 3D-printed bracket will fail under load. Each is a partial digital twin, and each reduces the number of costly surprises waiting on the workshop floor.

## TOOLS AS ENABLERS, NOT SOLUTIONS

Tools are seductive. A new project management platform, a better modeling environment, a more sophisticated version control system, each promises to improve how teams work. Some deliver on that promise. Many do not, because the tool is adopted without the process to support it.

The selection of tools should follow from the process, not precede it. A team that has chosen an iterative development approach needs tools that support rapid prototyping, continuous integration, and collaborative review. A team following a staged process needs tools that support traceability, formal documentation, and phase-gate reviews. Misalignment between process and tooling creates friction, workarounds, and ultimately, disillusionment.

Consider the experience of teams adopting model-based systems engineering tools during the transition to SysML v2. The language itself is more expressive, more precise, and more interoperable than its predecessor. But teams that adopted the tooling before establishing the practices, before agreeing on modeling conventions, before defining what would and would not be modeled, before training the team on the notation, found themselves spending more time managing the tool than managing the system. The tool became an obstacle dressed as an improvement.

The same pattern appears with artificial intelligence coding assistants. GitHub Copilot, Cursor, and similar tools can generate boilerplate code, suggest implementations, and accelerate development of unfamiliar functionality. A 2025 randomized controlled trial by METR found, however, that experienced open-source developers were nineteen percent slower when using AI tools on their own familiar repositories, despite believing the tools had made them twenty percent faster. The perception of acceleration did not match the reality of execution. AI coding tools are most valuable for exploration, for working in unfamiliar domains, and for automating the tedious. They are least valuable when they interrupt the flow of an expert who already knows precisely what to build. The lesson is not that AI tools are unhelpful. The lesson is that tools must be deployed where they create genuine leverage, not where they create the feeling of leverage.

For early-stage teams, simplicity is an asset. A well-maintained requirements table, a version-controlled repository, and a structured communication cadence often outperform expensive platforms that the team has neither the time nor the expertise to use effectively. Tools should scale with the team's needs, not the other way around.

## FABRICATION TOOLS AND RAPID PROTOTYPING

The democratization of fabrication has reshaped what is possible during Quest. Three decades ago, producing a functional prototype of a mechanical part required access to a machine shop, skilled machinists, and weeks of lead time. Today, a desktop 3D printer can produce a functional geometry overnight. Laser cutters, CNC routers, and PCB milling machines have followed similar trajectories from industrial exclusivity to workshop accessibility.

This accessibility carries both benefit and risk. The benefit is obvious: teams can iterate on physical form faster and more cheaply than ever. The risk is subtler. When prototypes are cheap, the temptation is to build before thinking, to use fabrication as a substitute for analysis rather than a complement to it. A team that prints twelve iterations of a bracket without first modeling its load path has not prototyped twelve designs. It has guessed twelve times.

The most effective teams use fabrication tools in dialogue with simulation. The digital model predicts. The physical prototype validates. The discrepancy between prediction and observation becomes the most valuable data point in the cycle, revealing where the model is incomplete and where reality behaves differently than assumed. This dialogue—digital prediction, physical validation, model refinement—is the engine of disciplined prototype development.

Additive manufacturing has also expanded into metals, composites, and even biological materials. SpaceX uses 3D-printed SuperDraco engine chambers in its Crew Dragon capsule, leveraging the geometric freedom of additive manufacturing to create cooling channels that conventional machining cannot produce. At the other end of the scale, bioprinting companies are fabricating tissue scaffolds and organ-on-chip platforms that allow pharmaceutical companies to test drug candidates without animal models. The tools have changed. The principle has not: fabrication serves learning, and learning serves decisions.

## CONFIGURATION MANAGEMENT: GOVERNING CHANGE

As prototypes evolve, the number of artifacts like code files, CAD models, test results, requirements documents, firmware images, grows faster than most teams anticipate. Without a system for tracking what changed, when, by whom, and why, the team loses the ability to reason about its own work. Configuration management provides that system.

At its simplest, configuration management ensures that what is being built matches what was designed, and that changes are deliberate rather than accidental. Version control for code, revision tracking for documents, and bill-of-materials management for physical assemblies are all expressions of the same underlying discipline: the discipline of knowing, at any moment, exactly what the product is.

This discipline matters more during Quest than many teams realize. A prototype that cannot be reproduced is a prototype that cannot be trusted. If the team cannot answer the question "What exactly did we test, and how does it differ from what we tested last week?" then the results of testing carry an uncertainty that no amount of analysis can resolve. Configuration management does not create rigor. It preserves rigor that has already been invested.

## THE PRODUCT SPECIFICATION

One of the key outputs of this phase is the product specification, a comprehensive document that captures what the product is, how it works, and what standards it must meet. The product specification integrates the system model, the product and technology requirements, the market requirements, and the design decisions made through trade studies and prototyping. It is the blueprint from which the product will be built, tested, and manufactured.

A well-crafted product specification does not merely describe the product; it explains it. It traces each design decision to a requirement, each requirement to a need, and each need to a user or market context. It anticipates questions that will arise during implementation and provides answers before they become blockers. Vague descriptions that sufficed during early prototyping, "use a strong adhesive," "ensure adequate clearance" must be replaced with quantified, verifiable statements. Every characteristic that can vary must have defined limits. Every process that can drift must have defined controls.

The product specification is the bridge between Quest and Accelerate. It captures what has been learned in a form that enables others, manufacturing engineers, test teams, regulatory specialists, supply chain partners to act on that learning without having been present for it. In this sense, the specification is not documentation. It is an act of translation, rendering hard-won understanding into a language the rest of the organization can use.

---

Process and tools are the infrastructure of execution. They do not create insight, but they preserve it. They do not generate quality, but they enable it. Choosing them well, deliberately, in alignment with the product and the team is the difference between organized progress and improvised activity. The team that selects its process with the same rigor it applies to its trade studies, and selects its tools with the same discipline it applies to its requirements, builds on a foundation that will hold. The team that inherits both by default builds on sand.

---

# 5
# INTELLECTUAL PROPERTY

Innovation that cannot be protected is innovation that can be copied. Intellectual property, the legal framework for protecting creations of the mind, is not an afterthought to product development. It is a strategic dimension that shapes what can be built, how it can be commercialized, and what value it can sustain over time. For many startups and technology companies, intellectual property is the most valuable asset on the balance sheet. A study by the United States Patent and Trademark Office found that startups with patent filings are more than twice as likely to secure funding and tend to raise larger rounds. Protected innovation is more valuable than unprotected innovation, not because legal documents create value, but because they prevent others from extracting it.

## THE FORMS OF INTELLECTUAL PROPERTY

Intellectual property rights are, at their core, a fence around creative work, a right to exclude others from using it. Each form of IP excludes others from different activities regarding different types of work. Patents protect inventions, novel, non-obvious, and useful products, processes, or compositions of matter, granting the holder a monopoly for typically twenty years in exchange for public disclosure. Trademarks protect brand identity, names, slogans, logos, and trade dress, indefinitely, as long as the mark remains in use. Copyrights protect original works of authorship, software code, documentation, artistic works, for the life of the author plus seventy years in most jurisdictions. Trade secrets protect confidential business information, formulas, algorithms, manufacturing processes, for as long as secrecy is maintained.

Each form requires different strategies for acquisition, maintenance, and enforcement. A comprehensive IP strategy considers all four dimensions, selecting the most appropriate protections for each aspect of the product. The choice between patents and trade secrets, for instance, is strategic rather than automatic. Patents are public and enforceable against anyone who independently discovers the same invention. Trade secrets are private, last indefinitely, and cover anything, but they offer no protection against reverse-engineering or independent discovery. The cubic equation remained a trade secret for decades in Renaissance Italy, conferring enormous advantage to the mathematicians who possessed it. A patent would have disclosed the method to the world but granted only a time-limited monopoly. The right choice depends on context: how easily the innovation can be reverse-engineered, how long the competitive advantage must last, and whether public disclosure serves the broader strategy.

## AI AND THE INVENTORSHIP QUESTION

Artificial intelligence has disrupted every dimension of intellectual property, but nowhere more consequentially than in the foundational question of who qualifies as an inventor. Computer scientist Stephen Thaler spent years attempting to name his AI system DABUS as the inventor on patent applications filed across multiple jurisdictions. South Africa granted the patent in 2020, the only country to do so. The United States, European Union, United Kingdom, and Australia all denied it. In the United States, the Federal Circuit ruled in Thaler v.

Vidal in 2022 that only natural persons can be inventors under the Patent Act.

The practical resolution came in November 2025, when the USPTO issued revised inventorship guidance that rescinded earlier, more ambiguous rules. The new guidance is unequivocal: AI systems are tools, analogous to laboratory equipment, software, or research databases. They may assist in the inventive process, but they cannot conceive inventions. The traditional legal test, whether a natural person formed a definite and permanent idea of the complete and operative invention, applies uniformly to all inventions, regardless of whether AI was involved. This means product development teams can patent AI-assisted inventions, but they must document how a human conceived each claim. Invention disclosure forms must identify only natural-person contributors. The lesson for teams using AI extensively in research and development is direct: meticulous documentation of human inventive thought is now an IP imperative, not an administrative nicety.

## AI AND THE AUTHORSHIP QUESTION

Copyright law has reached a parallel conclusion through a different path. Thaler also sought to register a copyright for a visual artwork his AI system generated autonomously. The Copyright Office denied the application, the district court upheld the denial, and in March 2025 the D.C. Circuit affirmed that the Copyright Act requires all eligible work to be authored in the first instance by a human being. In March 2026, the Supreme Court declined to hear Thaler's appeal, effectively settling the question for the foreseeable future: purely AI-generated works cannot be copyrighted under U.S. law.

The more nuanced question concerns AI-assisted works, where a human provides creative direction, selection, editing, or arrangement of AI-generated outputs. The Copyright Office has indicated that such works may be eligible for protection if the human contribution is sufficiently significant. The case of Allen v. Perlmutter, pending in Colorado, is testing exactly where that line falls. For product development teams, the practical implication is substantial: code generated entirely by AI tools like Copilot may not be copyrightable, which means competitors could freely use identical outputs. Teams should structure their AI workflows to ensure meaningful human creative contribution at every stage, and should document that contribution as carefully as they document inventive conception for patents.

The second front in AI copyright law concerns training data. In February 2025, a Delaware federal court ruled in Thomson Reuters v. ROSS Intelligence that using Westlaw's copyrighted legal headnotes to train a competing AI search engine was not fair use, the first major U.S. court decision on this question. Cases involving the New York Times against OpenAI, music publishers against Anthropic, and Getty Images against Stability AI remain pending, but the Thomson Reuters ruling signals that AI training on copyrighted content is not automatically protected. Product teams building AI systems must consider the provenance of their training data as a strategic and legal risk factor.

## TRADE SECRETS IN THE AI ERA

AI has simultaneously created new categories of trade secrets and new vectors for their destruction. Training datasets, model weights, fine-tuning configurations, and prompt engineering techniques all qualify as trade secrets, provided the organization maintains reasonable measures to keep them confidential. The challenge is that AI tools themselves can compromise that secrecy. In 2023, Samsung engineers inadvertently leaked proprietary source code by pasting it into ChatGPT to debug and optimize their work. The incident destroyed trade secret status for the disclosed material, because the information was transmitted to a third-party system without confidentiality protections. Samsung subsequently banned internal use of ChatGPT.

The Samsung incident is not an anomaly. It illustrates a structural vulnerability that every organization using AI tools must address. When employees paste proprietary code, customer data, or strategic documents into AI systems, they may be permanently destroying trade secret protection, regardless of whether the AI provider claims not to retain the data. Organizations must implement explicit AI usage policies that classify which information may and may not be shared with external AI tools, enforce access controls on sensitive data, and train employees to recognize the trade secret implications of their AI interactions. The tension between open-source AI models, which by definition cannot be trade secrets, and proprietary AI systems, which demand secrecy infrastructure, adds another dimension to the strategic calculation.

## IP STRATEGY AS PRODUCT STRATEGY

Intellectual property strategy should be developed alongside product strategy, not after it. This principle, always important, has become urgent in the AI era because the window between creation and public disclosure has collapsed. An algorithm shared in a conference paper, a model architecture described in a blog post, or a technique demonstrated in an open-source repository can be replicated worldwide within hours. The first-to-file patent system used in most countries means that a competitor who files before you, even if they invented later, may hold the rights. Alexander Graham Bell's telephone patent, filed before the invention was fully operational, transformed a scientific experiment into a global industry. The principle applies with even greater force today, when AI can accelerate both invention and imitation.

A freedom-to-operate analysis is equally essential. Before committing to a design, teams should understand whether their proposed approach, including any AI-generated components, infringes on existing patents or copyrights. AI systems may produce outputs that unwittingly incorporate elements from their training data, creating infringement risks that are invisible without deliberate analysis. AI is also transforming how this analysis is conducted: machine learning models now search patent and non-patent literature to find relevant prior art that traditional keyword searches miss, map competitive patent landscapes, and predict examiner behavior during prosecution. IP awareness should be cultivated throughout the team, not confined to a legal department. Engineers and designers are often the first to recognize novel methods or applications. When IP awareness is embedded in the culture of the team, opportunities for protection are identified early, when they are most valuable and least expensive to pursue.

Partnership agreements must now explicitly address AI-generated IP. When two organizations collaborate using AI tools, questions of ownership become complex: if a jointly trained model produces a novel solution, who owns the output? Contracts that do not specify how AI-generated innovations will be owned, licensed, and governed contain a structural gap, the kind of invisible architectural flaw this book has argued against from its opening pages.

---

> Intellectual property is the legal expression of innovation. In a world where AI accelerates both creation and copying, the fence around creative work matters more than ever. Teams that think about IP early, document human contributions systematically, and align their IP strategy with their product strategy are the teams that build enduring value. The invisible architecture of every successful product includes not just the technical decisions that shaped it, but the legal protections that sustain it.

---

# PREPARING FOR THE NEXT PHASE

The work of Quest is now complete. What began as a framed problem and a Proposal has been transformed into something far more substantial: a validated set of trade-offs, a system model that captures the product's architecture, requirements that encode both engineering precision and market intent, a development process chosen with care, and an intellectual property strategy that protects what has been built.

What comes next is Accelerate: the transition from validated design to production reality.

---

Quest was never about building a product. It was about building the understanding that makes building a product possible. That understanding is now in hand. The work of making continues.

---

# PART III
# ACCELERATE

## FROM VALIDATED LEARNING TO DISCIPLINED EXECUTION

1

# FROM LEARNING TO EXECUTION DISCIPLINE

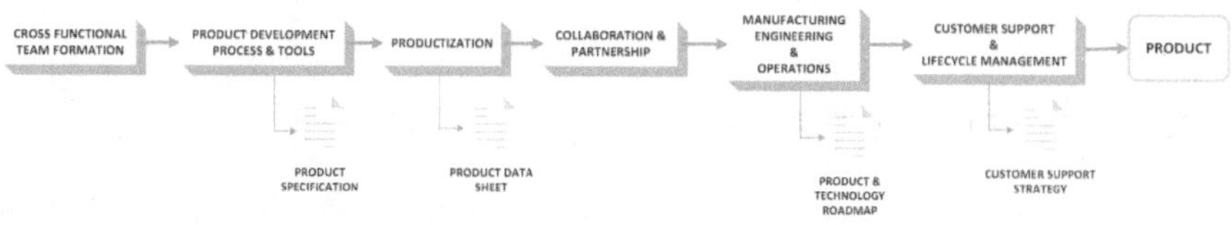


Figure 3: ACCELERATE FLOW

By the time a team reaches Accelerate, the nature of the work changes fundamentally. The primary question is no longer "What are we learning?" It becomes "What are we committing to?"

Accelerate is the phase where optionality begins to collapse by design. Decisions become binding. Interfaces harden. Dependencies multiply.

## THE SHIFT FROM VELOCITY TO COHERENCE

In Quest, speed is measured by learning velocity. In Accelerate, speed is measured by coherence. Moving fast while the system is unstable creates fragility. Moving deliberately while preserving system integrity creates leverage.

Consider the trajectory of SpaceX. The company's early Falcon 1 program was quintessential Quest work: rapid prototyping, public failures, fast iteration. But when SpaceX transitioned to the Falcon 9, the nature of the work shifted. The company committed to vertical integration on a scale the aerospace industry had never seen. That decision was a disciplined commitment that would eventually reduce launch costs by a factor of ten.

## BUILDING THE SYSTEM, NOT JUST THE PRODUCT

A defining feature of this phase is the expansion of scope. The product is no longer the sole center of gravity. The system that produces, delivers, supports, and evolves the product becomes equally important.

## SCALING IS A DESIGN PROBLEM

Scaling is often discussed as a growth problem. In reality, it is a design problem. Teams that succeed at scale rarely rely on brilliance. They rely on structures that allow ordinary decisions to produce extraordinary consistency.

---

> Accelerate is where ambition encounters manufacturing, markets, regulation, and time. It is where disciplined teams distinguish themselves, not by moving faster, but by moving together.

---

# 2
# FORMING THE CROSS-FUNCTIONAL TEAM

As products move from validated learning to execution, the team itself must be redesigned. The composition of a cross-functional team is not an administrative decision. It is a design decision.

## WHY COMPOSITION IS A DESIGN CHOICE

A product that involves embedded electronics, mechanical enclosures, software, and a cloud backend requires electrical engineers, mechanical engineers, computer scientists, and systems architects. The disciplines on the team must mirror the interfaces in the product system. It is not unusual that the people who end up in the Accelerate phase are different from those involved in the Quest phase. As an example, Philips Medical Systems chooses a completely different team in the final production than that used in the pre-production phase of the product. The reason is that this team needs to be intimately familiar with the production mechanisms. It uses a different cadence in the execution. The documentation approach is thorough and verification teams play a significant role.

## THE OBEYA PRINCIPLE: ALIGNMENT THROUGH PROXIMITY

In the early 1990s, Toyota embarked on designing the car of the twenty-first century. Takeshi Uchiyamada created what Toyota came to call the Obeya: a large, dedicated room where specialists from design, evaluation, production engineering, and manufacturing sat together and made decisions in real time. The Prius reached the market in under three years, more than a decade before any competitor.

The principle behind Obeya is not about physical space. It is about information visibility and decision proximity.

## SMALL TEAMS, CLEAR AUTHORITY

Lockheed's Skunk Works offers the complementary lesson. Clarence "Kelly" Johnson hand-picked twenty-three engineers. The XP-80 Shooting Star was delivered in 143 days, seven days ahead of schedule. Johnson's insight was that clarity of authority and minimal communication overhead allow talented people to make decisions at the speed the work requires.

## FROM TALENT TO STRUCTURE

Scale introduces a different requirement: clarity. Roles must be defined, ownership established, and interfaces stabilized. Execution discipline does not diminish creativity. It protects it by ensuring that effort compounds instead of colliding.

## BRIDGING ORGANIZATIONS

Many products in Accelerate involve collaboration between organizations. Shared artifacts, system models, requirement documents, and decision records become essential coordination mechanisms.

---

A team is not a collection of people with complementary skills. It is a system of relationships, authorities, and shared commitments.

---

# 3
# PRODUCT DEVELOPMENT AND PRODUCTIZATION

In 2013, a startup called Juicero raised over one hundred and twenty million dollars to build a countertop juice press. The engineering was, by most accounts, extraordinary: over four hundred custom parts, aircraft-grade aluminum, a pressing mechanism capable of exerting four tons of force. The prototype worked flawlessly. Investors were captivated. The technology performed exactly as designed.

Then Bloomberg reporters demonstrated that the proprietary juice packs could be squeezed by hand with comparable results. The entire value proposition collapsed. Juicero shut down within months.

Juicero did not fail because its engineers lacked talent. It failed because productization, the structured work of converting a working technology into something that can be sold, supported, and sustained, was never allowed to discipline design ambition. The product was over-engineered for its market. The manufacturing complexity was staggering. The unit cost was unsustainable. And no one with authority over these dimensions had been empowered to say so early enough to matter.

A working prototype is not a product. This distinction is the source of more failed ventures than any technical shortcoming. The distance between "something that works" and "something that can be sold, supported, and sustained" is vast, and it is populated by disciplines that most development teams encounter too late. Productization is the structured work of crossing that distance.

## THE FULL SCOPE OF PRODUCTIZATION

The breadth of productization surprises teams that have spent months focused on making the technology function. Quality management, user and service documentation, branding and merchandising, intellectual property filings, localization, pricing, packaging, sales literature, sales training, advertising, website and e-commerce support, customer support infrastructure, warranty management, and regulatory certification, each of these is a dimension that must be addressed before a product can be sustained in the marketplace. Neglecting any one of them does not merely create inefficiency. It creates a specific mode of failure.

This is why the role of product manager exists. The diversity and range of these dimensions extend so far beyond the physical implementation that it requires its own managerial attention. Only a product addressed by the full scope of productization can be truly sustained without undue downstream burden. The product manager orchestrates across all of them, not as a coordinator of tasks, but as a systems thinker who understands how pricing affects positioning, how packaging communicates brand, how documentation shapes the support burden, and how regulatory strategy constrains market entry timing.

## FROM "CAN WE MAKE IT WORK?" TO "CAN IT WORK ALWAYS?"

Early development rewards ingenuity. Engineers find ways around limitations. Processes are flexible. Decisions are reversible. This creativity is essential, and it does not scale.

Productization demands a different question: can this work predictably, repeatedly, and under variation? The shift affects every layer of the system. What was once implicit must become explicit.

What was once flexible must become intentional. If a design requires exceptional care to build correctly, it will eventually fail in the field.

James Dyson tested over five thousand prototypes of his cyclone vacuum before bringing it to market. Each prototype answered a question not about whether the technology worked, but about whether it could work under conditions real users would impose, dropped on tile floors, clogged with pet hair, operated by people who would never read the manual. The first iPhone went through dozens of iterations before its 2007 launch. Apple's obsession was not merely with functionality, but with the totality of the product experience: packaging that communicated quality before the device was touched, a supply chain capable of delivering millions of units within weeks, and a support infrastructure ready for a device category that had never existed.

## QUALITY AS FITNESS FOR PURPOSE

Quality assurance establishes the processes and controls that ensure repeatable, well-documented development, the Quality Management System that is often mandatory in regulated industries. Quality control then verifies that each manufactured instance faithfully replicates the assured design through measurement and statistical analysis. The distinction matters: quality assurance belongs to product engineering, designing products predisposed to meet requirements; quality control belongs to manufacturing, ensuring each copy matches the design. When statistical samples drift, the rate escalates to one hundred percent until the root cause is identified. The cost of building quality in is always less than the cost of failing to.

## BRAND, PACKAGING, AND THE CUSTOMER'S FIRST IMPRESSION

Products come and go, while brands remain. Volvo has accrued a promise of safety. BMW has branded itself as the Ultimate Driving Machine. The design language of a product, Apple's sharp, uninterrupted edges, BMW's Hofmeister kink defining the rear window pillar, carries across models, accessories, even the key in your pocket. Consistency of promise and fulfillment of promise are the two hallmarks of a successful brand.

Packaging is where brand meets physical reality. Apple's iPhone arrives in a tight-fitting box with sharp, well-defined edges, uncompromising orthogonality in flat, pure white. The molded fiber inserts, the paper origami structures wrapping cables, the absence of plastic, every element communicates that the attention in the packaging reflects attention in the product. When packaging is overlooked, it subtracts from potential rather than adding to it. Packaging defects contribute to over twenty percent of product recalls globally, representing not just financial loss but reputational erosion that compounds with every returned unit.

## LOCALIZATION, PRICING, AND THE PRODUCT IN THE WORLD

Consider the Sony Walkman, launched in 1979. The technology was not revolutionary; portable cassette players already existed. What Sony designed was a complete product system: the price point, the lightweight headphones, the name, and the distribution channels. The market requirements, portability, personal use, affordable pricing, cultural relevance, drove product decisions as much as any engineering specification. The Walkman succeeded because every dimension of productization was aligned with a single market insight.

Localization extends this principle to every geography. Language adaptation, character set support, regulatory variation, cultural norms, even the wall plug, each must be addressed. The key is architecting for localization from the start: externalizing all language and symbols from the product design so that a single architecture can serve Japanese, German, and Brazilian markets. Retrofitting localization after the fact is so burdensome that it becomes its own barrier to market entry.

Pricing is equally a productization discipline. The method of alternative minimum cost equates value to the least cost the customer would otherwise face. Target gross margin pricing sets a floor. Buyer behavior pricing steers customers toward preferred configurations. Reservation demand pricing, a low foundation with premium add-ons, has become the dominant subscription model. Each is a strategic choice, and the wrong pricing model can undermine a superior product as effectively as a technical deficiency.

## REGULATORY CERTIFICATION: THE GATE THAT CANNOT BE BYPASSED

Few products are free of regulatory influence, and the obligations multiply with every geography. The process of achieving compliance follows a consistent sequence: survey the applicable requirements, incorporate them into product requirements with full traceability, create test cases to verify compliance, conduct investigative testing, submit for independent third-party certification, file the necessary documentation, obtain clearance, and apply compliance marks to the product and its packaging. Items one through four in this sequence apply to all requirements, whether regulatory or otherwise, which simply underscores how integral compliance is with mainstream development. Teams that treat regulatory certification as an afterthought discover, painfully, that it has veto power over their launch date.

## CUSTOMER SUPPORT AND WARRANTY

After the initial euphoria of a purchase settles, the product must sustain a relationship. Customers need knowledge of product use. Defects must be handled through return merchandise authorization or, when widespread, through recalls. Damage requires repair infrastructure. And the warranty, a business decision at its core, must balance competitive positioning against the cost implied by failure rates and repair expenses. How a company responds when things go wrong often matters more to customers than how the product performs when things go right. Warranties can also cover preventive maintenance, limiting returns by keeping products in optimal condition. But they do not substitute for engineering quality into the product.

## THE PRODUCT SPECIFICATION AND THE DATA SHEET

Two documents crystallize the work of productization. The product specification captures what the product is as manufactured: every dimension, tolerance, material, and performance characteristic. Vague descriptions from prototyping, "use a strong adhesive," "ensure adequate clearance" must be replaced with quantified, verifiable statements. It is the contract between design and manufacturing. The product data sheet faces outward, translating technical capabilities into customer-facing language. Creating the data sheet forces the team to answer what the most important features are from the customer's perspective and what performance claims can be substantiated. If these questions cannot be answered clearly, the product is not yet ready for market.

---

> Productization is not the final step of development. It is the first step of the product's life in the world. The valley of death between prototype and market is not a chasm to be leapt across with courage alone. It is a landscape to be navigated with structure, discipline, and systems thinking. Productization is where the team's promises become the customer's experience, the craft of making things work, not once under ideal conditions, but always, under real ones.

---

# 4
# MANUFACTURING PLANNING AND OPERATIONS

## FROM DESIGN INTENT TO PRODUCTION REALITY

A product that cannot be manufactured reliably is not a product. It is a demonstration. Yet manufacturing is where many promising innovations die, not because the design was flawed, but because the system required to produce it was never designed at all.

The pattern is remarkably consistent. A team builds a working prototype. It performs beautifully in the lab. Investors are impressed. Customers express interest. Then the question shifts from "Can we make this work?" to "Can we make ten thousand of these, reliably, at a cost someone will pay?" The silence that follows is the sound of a gap that should have been closed months earlier.

Consider Juicero, the Silicon Valley startup that raised over $120 million to build a countertop juice press. The product worked. The engineering was, by most accounts, sophisticated, over four hundred custom parts, aircraft-grade aluminum, a pressing mechanism capable of exerting four tons of force. But the manufacturing complexity was staggering, the unit cost unsustainable, and when Bloomberg reporters demonstrated that the proprietary juice packs could be squeezed by hand with comparable results, the entire value proposition collapsed.

Juicero did not fail because its engineers lacked talent. It failed because manufacturing reality was never allowed to discipline design ambition. The product was over-engineered for its market, and no one with manufacturing authority had been empowered to say so early enough to matter.

The lesson applies broadly. Manufacturing planning is not a downstream activity to be handled after the real work of design is complete. It is a concurrent discipline that shapes what is possible, what is economical, and what will endure.

## WHY MANUFACTURING IS A DESIGN PROBLEM

The essential insight is structural: manufacturing is not the execution of design. It is a co-equal domain that constrains, informs, and often improves design. The relationship between the two is not sequential but iterative. A product designed without manufacturing input will be redesigned, expensively, when it reaches the factory floor. A manufacturing process designed without understanding product intent will produce parts that meet tolerances but miss the point.

Three questions define the manufacturing system. Product engineering answers what to make. Manufacturing engineering answers how to make it. Manufacturing operations answer how many and when. These are not handoffs. They are interdependent decisions that must be made in dialogue.

The consequences of separating them are well-documented. When Apple launched the Mac Pro in 2013, the sleek, cylindrical "trash can" design and the thermal architecture were so tightly constrained by the form factor that Apple could not upgrade the GPU without a complete redesign. The manufacturing process had been optimized for a product that left no room to evolve. Apple's own executives later acknowledged the design had painted them into a corner, and the product languished for six years before being replaced by a modular tower that prioritized upgradeability over aesthetics. Manufacturing constraints, had they been surfaced earlier, would have reshaped the architecture before it hardened.

SpaceX offers the counter-example. When engineers at SpaceX proposed using friction stir welding for the Falcon 9's fuel tanks, the decision was not made by the design team alone. Manufacturing engineers were in the room, evaluating whether the process could achieve the required joint strength at production rates, whether the tooling could be built in-house, and whether the technique would scale as launch cadence increased. The manufacturing question was not "Can you build what we designed?" It was "What should we design, given what we can build?" That inversion, manufacturing as input rather than recipient, is what enabled SpaceX to produce rockets at a pace the industry had never seen.

## WHAT MANUFACTURING PLANNING MUST ADDRESS

Every discrete product exists as a hierarchy of assemblies, sub-assemblies, and components. Each level requires its own production process, and the interactions between those processes determine whether the whole system coheres or fractures. Manufacturing planning must address several dimensions simultaneously.

Quality must be engineered into the process, not inspected into the product. The distinction between quality assurance and quality control matters here. Quality assurance belongs to product engineering, designing products predisposed to meet requirements. Quality control belongs to manufacturing engineering, specifying the means to ensure each instance matches the design. Statistical process control, measuring the parameters that influence output and adjusting processes to reduce sensitivity to variation, is what makes quality systemic rather than heroic.

The choice of production model shapes everything downstream. The infrastructure required for full custom products, bespoke, high-touch, built start-to-finish by a dedicated team, bears almost no resemblance to what is needed for high-volume consumer goods, where micro-staged, high-speed machinery tolerates no variation and diverts non-conforming units automatically. Between these extremes lie high-custom order-to-cash systems with parametrized designs and reprogrammable lines, small-batch production with flexible cellular manufacturing, and semi-custom platforms like automobiles that mix high commonality with extensive optionality. Each model makes different demands on capital, workforce skill, IT systems, and organizational structure. Choosing the wrong model, or, more commonly, choosing none deliberately, is a strategic error that compounds daily.

## OPERATIONS: WHERE PLANS MEET REALITY

Manufacturing operations orchestrates execution against fluctuating demand, supplier variability, and the constraints of physical infrastructure. The information backbone, whether a standalone MRP system handling scheduling and inventory, or a fully integrated ERP platform spanning finance, supply chain, and customer management, determines how effectively the organization can respond when reality departs from plan. And reality always departs from plan.

Toyota understood this before most. When the 2011 Tohoku earthquake and tsunami devastated Japan's northeast coast, Toyota's supply chain was severely disrupted. But Toyota recovered faster than any competitor, not because it was lucky, but because decades of investment in supplier relationships, real-time visibility into its multi-tier supply chain, and a culture of rapid problem-solving had created resilience as a system property. Toyota knew, within days, which of its thousands of suppliers were affected, which components were at risk, and which alternative sources could be activated. Other automakers, with less visibility into their own supply networks, took months longer to recover. The difference was not resources. It was information architecture and relationship design.

The emerging role of artificial intelligence in manufacturing operations is accelerating this dynamic. Predictive maintenance systems that detect tool wear before failure, adaptive scheduling algorithms that rebalance production in real time, and quality optimization models that correlate process parameters with output variation are not speculative, they are deployed capabilities reshaping what operations can achieve. But they depend on infrastructure decisions made today: sensor investment, data architecture, communication resilience. Teams that defer these decisions will find themselves locked out of capabilities their competitors take for granted.

## THE OUTSOURCING DECISION

For most startups and early-stage ventures, unless the manufacturing process itself is where intellectual property resides, outsourcing should be the primary approach. Capital is finite. Manufacturing expertise takes years to develop. The opportunity cost of building production capability before the product-market fit is validated is severe.

But outsourcing is not abdication. It is a system design problem, one that demands the same rigor applied to any other interface in the product system. Partner selection, contract terms, IP protection, quality requirements, and logistics coordination must all be designed explicitly. The Boeing 787 lesson, discussed elsewhere in this book, remains the definitive cautionary tale: outsourcing at unprecedented scale without designing the organizational interfaces that outsourcing demands produced coordination breakdown, quality failures, and cost overruns measured in billions.

The decision to bring manufacturing in-house as the business grows is itself a trade study, informed by the same principles of structured evaluation that govern every other consequential choice in product development. When Dyson moved vacuum cleaner production from the UK to Malaysia in 2002, the decision was driven not by labor cost alone but by proximity to suppliers, access to a growing Asian market, and the ability to co-locate engineering and manufacturing teams that could iterate on production processes in real time. The manufacturing location was a strategic choice, not an administrative one.

## THE INTEGRATION IMPERATIVE

The most consequential insight about manufacturing planning and operations is simple to state and difficult to practice: it must be integrated into product development from the beginning. Not consulted. Not informed. Integrated, with authority to shape decisions, veto designs that cannot be produced economically, and surface constraints that the design team cannot see from where it sits.

The organizations that do this well, Toyota with its Obeya rooms, SpaceX with its vertically integrated engineering-manufacturing feedback loops, and Apple when it gets the balance right, treat manufacturing not as a cost center but as a strategic capability. The organizations that do not find themselves redesigning products on the factory floor, negotiating quality compromises under schedule pressure, and discovering supply chain vulnerabilities at the worst possible moment.

---

> Manufacturing is where intent becomes tangible. It is where the abstract commitments of design encounter the concrete realities of material, process, and variation. A product that cannot be made reliably, at cost, and at quality is not yet a product. It is a promise waiting to be kept.

---

# 5
# COLLABORATION, PARTNERSHIPS, AND THE EXTENDED SYSTEM

No product reaches scale alone. By the time a product enters the later stages of Accelerate, its success depends on coordination across organizational boundaries, suppliers, manufacturing partners, distributors, regulators, and support organizations all become part of the system. This is where many well-designed products falter, not because the product is flawed, but because the system around it is under designed.

## PARTNERS AS SYSTEM COMPONENTS

In Accelerate, external entities become system components, with interfaces, requirements, and dependencies that must be designed as deliberately as any internal subsystem. The decision to collaborate or partner is a response to specific barriers and gaps: entry barriers such as regulatory burdens and intellectual property constraints, technology gaps, capacity gaps, geographic gaps, and distribution channel gaps. The first imperative is to choose strategies that involve the fewest such barriers. When the strategy remains compelling despite them, a collaboration or partnership with a gap-filling entity becomes the rational response.

The choice between a simple buyer-vendor arrangement and a deeper partnership depends on two variables: the degree of control required and the anticipated duration. Collaborations sit at one end of the spectrum, lower commitment, shorter term, reversible. Partnerships sit at the other, higher commitment, longer duration, more control, with the possibility of eventual acquisition. Understanding where a given relationship belongs on this spectrum is itself a design decision.

## THE BOEING 787 REVISITED

Boeing's outsourcing strategy for the 787 was not inherently flawed. The vision was sound: a revolutionary aircraft built from composite materials, with seventy percent of design and manufacturing distributed to over fifty partners worldwide. The error was treating organizational interfaces as administrative rather than engineering problems. The depth of coordination required, constant alignment on technical details, real-time resolution of integration issues, management of sub-tier suppliers across five hundred organizations in ten countries, was not designed into the system.

## DESIGNING FOR COLLABORATION

Collaboration at scale must be designed, not assumed. Kelly Johnson's Skunk Works understood this. His partnerships with Pratt & Whitney and Shell were built on clear specifications, direct communication, and mutual accountability. Johnson demanded close cooperation and liaison on a day-to-day basis. He understood that trust is not a substitute for structure, it is the product of structure.

The spectrum of arrangements is wide. Informal collaborations work where benefits are self-evident and conflicts absent. Joint agreements, development, marketing, technology-sharing, provide structure where conflicts exist. Coopetition, where competitors collaborate on shared challenges while competing in the marketplace, requires careful design: airline codesharing and the standards consortia behind Bluetooth, USB, and IEEE 802.11 are examples. Full joint ventures, capitalized as new entities with shared equity, represent the deepest commitment, are difficult to unwind, and often evolve from prior collaboration that proved its value.

## AI AND THE TRANSFORMATION OF COLLABORATION

Every dimension of collaboration described above is now being reshaped by artificial intelligence. The transformation is not speculative. It is operational and accelerating.

Partner discovery, traditionally a months-long process of fragmented spreadsheets and serendipitous introductions, is being compressed by AI platforms that scan thousands of companies, patents, and publications to identify partners based on capability matching and strategic fit. These systems assess financial health, cultural alignment, and technology maturity using natural language processing of public filings and social signals. Research suggests AI-driven discovery reduces time-to-market by over thirty percent. Critically, modern approaches start with desired outcomes, new markets, revenue acceleration, and work backward to identify which partners best fill specific gaps.

Due diligence has been similarly transformed. AI-powered patent analytics map the intellectual property landscape around a partnership domain, identifying freedom-to-operate risks before negotiations begin. Supply chain risk modeling, the kind of analysis that might have flagged the 787's vulnerabilities, can now predict partner dependencies and geopolitical exposure with a sophistication unavailable a decade ago. And collaboration execution may be the most consequential area of change: AI agents coordinate cross-organizational workflows, flag schedule risks, and identify interface conflicts in real time.

This amounts to a digital Obeya, extending Toyota's principle of information visibility and decision proximity across organizational and geographic boundaries, with shared dashboards and AI-curated project status replacing physical co-location. The principle endures; its reach has expanded dramatically.

## WHEN AI REPLACES THE PARTNER

A more disruptive question is emerging: when does an AI tool replace the need for a traditional partnership entirely? AI translation tools now substitute for local language partners in certain market entry scenarios. Domain-specific models fill technology gaps that previously required specialized subcontractors. Microsoft's Discovery platform demonstrated in 2025 that agentic AI could compress research timelines from months to hours, reducing dependence on academic and R&D partnerships.

This does not make partnerships obsolete. It redefines the barriers and gaps that drive partnership decisions. A technology gap that once required a joint development agreement may now be fillable with an AI tool and a licensing fee. Teams must evaluate whether each gap is best addressed by a human partner, an AI system, or a combination of both.

## INTELLECTUAL PROPERTY IN AI-MEDIATED PARTNERSHIPS

When AI tools contribute to joint inventions, traditional ownership frameworks strain. If two organizations share data to train a model that produces a novel solution, who owns the output? AI-assisted IP tracking tools now document contribution lineage across organizations, and AI systems monitor for potential infringement, detecting patent proximity issues before they become disputes. Partnership agreements must now explicitly address AI-generated IP. Any collaboration agreement that does not specify how AI-generated innovations will be owned, licensed, and governed contains a structural gap, the kind of invisible architectural flaw this book has argued against from its opening pages.

## NEW ANTI-TRUST CONSIDERATIONS

AI-mediated collaboration introduces anti-trust risks that traditional frameworks are only beginning to address. Algorithmic collusion, where AI pricing systems converge on supercompetitive prices without explicit human agreement, is the most consequential. The Department of Justice's 2024 lawsuit against RealPage alleged that the company's algorithmic pricing tool enabled landlords to coordinate rents across markets. Settlements have exceeded one hundred million dollars. California, New York, and Connecticut have enacted laws restricting algorithmic price-fixing, and the EU is considering classifying AI businesses as gatekeepers under an expanded Digital Markets Act.

For product development teams, the implications are direct. Data-sharing partnerships with competitors require explicitly defined anti-trust boundaries. AI tools processing competitively sensitive information need the same scrutiny as human conversations between competitors.

## NEW MODELS OF COLLABORATION

Beyond AI, the structure of collaboration itself is evolving. Platform companies, Apple, Salesforce, AWS, have created partnership ecosystems that differ fundamentally from traditional bilateral arrangements. Participation is governed by platform rules, API specifications, and marketplace dynamics rather than negotiated contracts. The shift from negotiated partnerships to ecosystem participation represents a category of inter-organizational relationship that the traditional collaboration-partnering spectrum does not fully capture. Organizations must now consider ecosystem strategy as a distinct discipline: not merely who to partner with, but which platforms to build on.

Digital twin technology enables another new model. Shared digital twin environments allow organizations to collaborate on product development by simulating subsystem interactions, without sharing underlying proprietary design details. Partners can test integration virtually before committing to manufacturing, resolving one of the fundamental tensions in technical collaboration: the need for deep integration with the imperative to protect proprietary knowledge.

Sustainability mandates are creating an entirely new category of required partnerships. Scope 3 emissions reporting demands supply chain transparency that only collaborative data sharing can provide. Circular economy requirements, product take-back, recycling, remanufacturing, demand partnerships that did not exist a decade ago. Competitors in the automotive industry now collaborate on shared battery recycling infrastructure, driven not by strategic choice but by regulatory imperative. These are not optional alliances. They are system requirements imposed by the operating environment.

## MANUFACTURING, SUPPLY, AND THE ROADMAP

Manufacturing and supply chains are strategic capabilities, not cost centers. The manufacturing engineer who identifies an unreliable tolerance is providing design input. The supply chain analyst who flags a single-source dependency is identifying a system vulnerability. Recent history, semiconductor shortages, pandemic disruptions, has made this vivid. Robust systems are designed with slack, alternatives, and transparency.

Collaboration and partnership decisions must be informed by a longer view. The product and technology roadmap bridges execution demands with the trajectory of the product family and partnership ecosystem. A roadmap is not a prediction. It is a declaration of intent, essential for aligning partners, prioritizing investments, and sequencing capabilities.

---

> Collaboration is not a soft skill bolted onto hard engineering. It is a system design discipline. When partnerships are designed with the same rigor as technical interfaces, with clear specifications, shared models, explicit accountability, and an honest assessment of where AI can augment or replace traditional arrangements,

they become sources of capability rather than sources of risk. The invisible architecture of every successful product includes the relationships that built it.

---

# 6
# CUSTOMER SUPPORT AND LIFECYCLE MANAGEMENT

A product's success is not measured at launch. It is measured across its lifetime, through updates, market shifts, regulatory changes, component obsolescence, customer evolution, and eventual succession. Lifecycle management is the discipline of designing for this entire arc, not merely the first act. Customer support is the instrument through which that arc is sustained. Together, they form the final layer of the invisible architecture: the systems that determine whether a product endures or erodes after it enters the world.

These disciplines are not afterthoughts. They are design inputs that must be considered during Accelerate, when architectural decisions still have leverage. A product designed without considering how it will be supported, updated, and eventually replaced is a product optimized for a moment that has already begun to pass.

## DESIGNING FOR EVOLUTION

Every product exists in a changing environment. Technologies advance. Components reach end of life. Regulations shift. Customer expectations evolve. Competitors respond. A product designed without consideration for these realities is optimized for a snapshot, not a trajectory.

Lifecycle management begins not after launch, but during Accelerate. Decisions made about architecture, modularity, interface stability, and documentation determine how gracefully the product will adapt to change. A monolithic architecture that cannot be updated without wholesale redesign becomes a liability. A modular architecture with well-defined interfaces allows components to evolve independently, extending the product's useful life and reducing the cost of adaptation.

Tesla's over-the-air update model illustrates what becomes possible when evolution is designed into the architecture from the start. Tesla vehicles receive software updates that fix defects remotely, add entirely new features, and improve performance, all without the customer visiting a service center. A car purchased in 2022 may drive differently, and better, in 2025. This blurs the traditional boundary between product generations. The product improves after purchase, and the lifecycle extends not through physical intervention but through digital evolution. The architectural decision that made this possible was not made at the point of update. It was made years earlier, when the vehicle's electronic architecture was designed to support continuous software deployment.

## CUSTOMER SUPPORT AS PRODUCT DESIGN

Support is not an after-sales function. It is part of the product experience. How a company responds when things go wrong often matters more to customers than how the product performs when things go right. A customer support strategy, developed during Accelerate, defines how issues are detected, escalated, and resolved. It specifies what diagnostic information the product must provide, what documentation must exist, and what organizational capabilities must be in place to sustain the product in the field.

Effective customer support rests on three pillars: people, process, and technology. The human element remains foundational, empathetic, knowledgeable, action-oriented agents who acknowledge the customer's reality, listen carefully, own the issue, and follow through to resolution. The most common support failures are not technical. They are human: failure to acknowledge frustration, failure to listen, lacking empathy, poor follow-through, and hiding behind automation to avoid personal interaction. These failures erode trust faster than any product defect.

Process ensures coherence. Few things alienate customers more completely than a chaotic process that forces repeated explanations, loses prior records, and fails to keep promises. Technology provides the infrastructure, systems to manage customer data, record and track issues to resolution, and surface patterns that inform product improvement. When these three elements are aligned, every customer interaction becomes an opportunity to strengthen the relationship. When any one is missing, even excellent performance in the others cannot compensate.

## THE AI-AUGMENTED SUPPORT INFRASTRUCTURE

Artificial intelligence has fundamentally reshaped the support infrastructure. The shift is not merely from human agents to automated chatbots. It is from deflection, discouraging customers from reaching a human, to resolution: AI systems that perform tasks, execute workflows, and solve problems end-to-end. Modern AI agents connect via APIs to customer relationship management systems, order management platforms, and payment gateways. They do not merely answer questions. They issue refunds, update accounts, and process returns.

Klarna's experience offers the most instructive case study of this transition, and its limits. In early 2024, the Swedish fintech deployed an OpenAI-powered assistant that handled 2.3 million customer service conversations in its first month, performing the equivalent work of seven hundred full-time agents. The company projected a forty-million-dollar profit improvement. Resolution times dropped to approximately two minutes, and repeat contacts fell by twenty-five percent. By any operational metric, the deployment was a success.

Then the cracks appeared. By mid-2025, CEO Sebastian Siemiatkowski acknowledged that cost had been the predominant evaluation factor, resulting in lower quality. Customers encountered AI responses that sounded generic rather than brand-specific. Complex issues were handled inadequately. The company pivoted to a hybrid model, rehiring skilled human agents for complex and emotionally sensitive interactions while expanding AI to handle routine queries and even some second-tier support. The lesson Klarna learned, at significant cost, aligns precisely with the principle this book has emphasized from its opening pages: the invisible architecture matters. An AI deployment optimized solely for cost reduction, without designing the organizational structure around it, produces the same category of failure as any other architectural shortcut.

The most effective model emerging from this experience is what practitioners call the copilot approach. Rather than replacing human agents, AI augments them, providing real-time knowledge retrieval from product documentation, suggesting responses based on conversation context, automatically summarizing customer history so no one repeats their problem, and detecting emotional distress through sentiment analysis to trigger proactive escalation. The agent copilot preserves the human empathy that the best support organizations depend on while eliminating the inefficiencies that make support expensive and inconsistent.

## PREDICTIVE SUPPORT AND THE DIGITAL TWIN

The most consequential shift in customer support is from reactive to predictive. Traditional support waits for something to break. Predictive support detects anomalies and anticipates failures before the customer experiences them. This transforms support from a cost center into a value-creation engine.

Rolls-Royce's TotalCare program represents the mature expression of this principle. Under TotalCare, airlines do not purchase jet engines. They purchase thrust, paying per flying hour while Rolls-Royce retains ownership and responsibility for maintenance. Digital twins of each engine receive continuous telemetry data from thousands of sensors, enabling real-time health monitoring and predictive maintenance that prevents aircraft-on-ground events worth millions of dollars per hour of delay. The digital twin bridges design intent and field reality, feeding operational data back into engineering to inform the next generation of engines. Support, in this model, is not a post-sale activity. It is the product.

The digital twin market reflects this trajectory. Projected to grow from approximately twenty-five billion dollars in 2025 to over two hundred and fifty billion by 2032, digital twins are moving from aerospace and defense into automotive, industrial automation, and consumer technology. A 2025 Hexagon survey found that ninety-two percent of companies deploying digital twins reported returns above ten percent, with over half reporting at least twenty percent return on investment. For customer support organizations, digital twins mean that agents can see the customer's product state in real time, diagnosing issues remotely before dispatching repair, and in many cases resolving problems before the customer knows they exist.

## CLOSING THE LOOP: FROM SUPPORT DATA TO PRODUCT INTELLIGENCE

When customer support data is aggregated and analyzed systematically, it becomes one of the most powerful inputs to product development. Every complaint, every support ticket, every social media mention contains information about how the product performs in reality, as opposed to how it was designed to perform in theory. The gap between these two is where the next generation of the product lives.

Natural language processing and sentiment analysis now enable organizations to mine every interaction, support calls, chat transcripts, online reviews, community forums to surface patterns that human analysis would miss. Systemic product issues emerge from thousands of individually minor complaints. Unmet needs surface from unstructured feedback. Churn prediction models identify at-risk customers based on support interaction patterns, enabling intervention before the customer leaves. This intelligence flows back into the Spark phase of the next product cycle, informing problem framing and opportunity mapping with field evidence rather than assumption.

The organizations that master this feedback loop gain a compounding advantage. Each product generation benefits from the accumulated field intelligence of its predecessors. When executed well, every customer issue becomes an asset whose value multiplies when aggregated across the installed base. This is the deepest justification for investing in customer support infrastructure: not merely to retain customers, but to accelerate the learning velocity of the entire product development system.

## SUSTAINABILITY, REGULATION, AND THE CHANGING LIFECYCLE

The regulatory landscape is reshaping lifecycle management in ways that product teams can no longer treat as peripheral. The European Union's Ecodesign for Sustainable Products Regulation, which entered force in 2024, introduces the Digital Product Passport, a mandatory digital record that every regulated product must carry, containing material composition, carbon footprint, repair instructions, and recycling pathways. Batteries must comply by February 2027, with textiles, electronics, steel, and furniture following by 2030. The Digital Product Passport is not a label. It is an infrastructure requirement that demands lifecycle data management from design through end-of-life.

The EU Right to Repair Directive compounds this shift, requiring manufacturers to make repair information, spare parts, and diagnostic tools available to consumers and independent repair providers. Together with the Cyber Resilience Act, which mandates security patching throughout a connected product's lifecycle, these regulations create a new category of obligations that extend far beyond the point of sale. End-of-life no longer ends the manufacturer's responsibility. The 2016 Mirai botnet attack, which exploited end-of-life IoT devices with default passwords to disable major internet infrastructure, demonstrated what happens when lifecycle security obligations are ignored.

For product development teams, the implication is direct. Sustainability, repairability, and security are no longer optional features. They are architectural requirements that must be designed in during Accelerate. Companies that view these regulations as compliance burdens will find themselves perpetually catching up. Companies that treat them as design inputs, building modular, repairable, updatable products from the start, will discover that the same architecture that satisfies regulators also satisfies customers and extends product life. Apple's Daisy robot, which disassembles two hundred iPhones per hour for material recovery, and Caterpillar's remanufacturing program, which returns heavy equipment components to like-new condition at fifty to sixty percent of new cost, demonstrate that circular economy practices can be both environmentally responsible and economically compelling.

## PLANNING FOR SUCCESSION

Every product, no matter how successful, eventually faces a transition: reinvention, replacement, or retirement. Lifecycle management treats this inevitability not as a distant concern but as a present design condition. What comes after this product? What technologies will it feed? What platforms will it establish? What knowledge will it generate?

The shift from product ownership to product-as-a-service models is changing the economics of succession. When Rolls-Royce sells thrust rather than engines, and John Deere sells precision agriculture data services alongside tractors, the manufacturer's incentives align with the customer's interest in product longevity. Planned obsolescence, once a standard lifecycle strategy, gives way to planned longevity when revenue depends on continuous use rather than replacement. Subscription models do not eliminate the need for succession planning. They change its character: from managing a transition between discrete product generations to orchestrating the continuous evolution of a platform.

The highest form of lifecycle success is not a product that lasts forever, but a system that enables its successor. Products that are well understood, well documented, and well governed provide a foundation for the next generation of innovation. The accumulated field data from a well-instrumented product, flowing through digital twins and AI analytics, becomes the most valuable input to the next cycle of Spark.

---

A product's launch is not its culmination. It is the beginning of a relationship, with customers, with markets, with time itself. Lifecycle management and customer support are the disciplines that sustain that relationship. When they are designed with the same rigor as the product's technical architecture, when the invisible infrastructure of support, feedback, and evolution is treated as a system to be engineered rather than a cost to be minimized, products do not merely endure. They improve. And the organizations that build them learn faster than those that do not.

---

# CLOSING: EXECUTION AS SYSTEM DESIGN

This book began with a building that collapsed, not because its walls were weak, but because its foundation was never examined.

Better Place raised eight hundred and fifty million dollars, deployed battery-swapping stations that worked flawlessly, and partnered with one of the world's largest automakers. The technology performed exactly as designed. And by 2013, the company's assets were liquidated for less than half a million dollars. The engineering was brilliant. The invisible architecture was fatally incomplete.

Across these chapters, a single pattern has emerged: execution failures are almost always organizational failures disguised as technical ones.

The Boeing 787 stumbled because organizational interfaces were under designed. The Skunk Works succeeded because the organizational structure allowed brilliant people to do their best work without obstruction. Toyota's Prius reached market a decade ahead of competitors because the Obeya system enabled real-time, cross-functional decision-making. SpaceX reduced launch costs by a factor of ten by redesigning the execution system itself.

In every case, the organization's architecture determined the product's architecture.

## THE ARC OF THE JOURNEY

*Spark established the discipline of seeing before acting.*
*Quest transformed that clarity into disciplined learning.*
*Accelerate converted that understanding into commitment.*

This progression from seeing to learning to building is not a formula. It is a discipline. And disciplines, unlike formulas, improve with practice.

## THE VISIBLE ARCHITECTURE

This book has been an effort to make the invisible architecture of product development visible, to render the implicit explicit, to surface the upstream decisions that determine whether what follows will succeed or struggle.

> Execution excellence is not the absence of problems. It is the presence of systems that detect problems early, resolve them efficiently, and learn from them continuously.

The odds against any new product remain severe. This book does not change that reality. What it offers is a systematic way to shift those odds, not through luck or genius, but through disciplined thinking applied at the moments where it matters most.

What was missing was never creativity. It was the structured understanding of what it actually takes to bring a product from concept to commercialization, the invisible architecture that supports everything the world eventually sees.

> This book has made that architecture visible. What you build with it is yours.

> When the right approach is chosen early, progress compounds.

When it is not, even flawless execution struggles to recover lost ground. The work began here. What matters is what follows.

# ABOUT THE AUTHOR

Imran H. Khan has nearly five decades of engineering experience, including technology development at MIT Lincoln Laboratory and Draper Laboratory. His career spans defense, aerospace, and advanced technology development, with deep expertise in systems engineering, embedded systems, computer vision, and AI integration.

Through NovaTechX, he leads Fourth Industrial Revolution education programs that bridge the gap between academic theory and industry practice, operating in Pakistan and expanding into economically under-served regions of the United States.

**Connect with the Author:**
Website: novatechx.io
LinkedIn: linkedin.com/in/imranhkhan/

**Continue the Journey**
If this book has helped you, please consider leaving a review on Amazon. Your feedback helps other readers discover this work.

www.ingramcontent.com/pod-product-compliance
Lightning Source LLC
LaVergne TN
LVHW100328210926
848480LV00003B/1290